The Magical Journey of Monarch Butterflies

By the Nature Narrators:

Eden Alvarez, Gracelyn Bowers, and Lilah Cox

with guidance by Janet Johnson and Jackie Alvarez

Copyright © 2024 Janet Johnson

All rights reserved.

ISBN: 9798323208227

Dedication

This book is lovingly dedicated to Carol Esarey, Lilah's great grandma, whose garden is a sanctuary for monarch butterflies. Carol's dedication to raising monarchs and cultivating milkweed has nurtured more than just the flora and fauna under her care—it has inspired a legacy of conservation and wonder in the hearts of her family and community. Through her eyes, we've learned to see the magic in the minutiae of nature and the importance of each creature's role in our world. Thank you, Carol, for being our guide on this magical journey. May this book carry your spirit of curiosity and conservation to new generations eager to explore the wonders of nature.

Forward

Welcome to an exciting journey into the world of monarch butterflies and their indispensable companion, milkweed. This book is not just a collection of activities; it is an invitation for elementary school children to dive into the realms of science and the environment, while nurturing their creative writing skills and having fun along the way.

Monarch butterflies, known for their striking orange and black wings and remarkable migration patterns, serve as a perfect entry point for young learners to explore broader ecological concepts and the importance of biodiversity. Milkweed, the sole food source for monarch caterpillars, introduces children to the concept of plant-animal interactions and the delicate balance of our ecosystems.

Each chapter in this book has been crafted with the enthusiasm and curiosity of three bright young minds: Lilah, Gracelyn, and Eden. These three elementary school girls not only contributed their ideas but also engaged in research, bringing authenticity and a child's perspective to each activity. Their involvement exemplifies the potential of young learners to contribute meaningfully to educational content.

While some adults might question the inclusion of advanced vocabulary in a children's book, we believe in challenging young minds. Just as children decipher complex information on Pokémon cards, they can grasp sophisticated scientific terms, which are explained clearly and integrated into fun activities.

Whether used as a standalone resource or as an introduction to raising monarch butterflies from eggs, the activities in this book promise a rich, hands-on learning experience. It's a chance for children to see themselves as scientists and storytellers, discovering the world through the wonder of monarchs and milkweed.

Enjoy this adventure into science, storytelling, and discovery. May it spark curiosity and inspire a deeper appreciation for the natural world among its young readers.

Contents

Welcome to the Monarch Butterfly Adventure! 1

Design a Pretend Milkweed 5

Predators 8

Milkweed Safety 13

Parts of the Milkweed Plant 20

Time to Be a Plant Scientist with Milkweed! 24

The Amazing Lifecycle of a Monarch Caterpillar 28

Identifying the Gender of Monarch Butterflies 32

A Fluttering Legacy: The Monarch's Gift from Great Grandma Carol .34

Welcome to the Monarch Butterfly Adventure!

Hello future scientists and nature explorers! You're about to embark on an exciting journey with one of nature's most fascinating and beautiful creatures - the **monarch** butterfly. This workbook is your guide as you delve into the world of monarchs, from tiny eggs to fluttering butterflies.

Why Monarchs? Monarchs are more than just pretty faces with wings. They play a vital role in our **ecosystem**, particularly in pollinating plants. However, their numbers have been declining, and they need our help! By participating in this project, you are not just learning; you're also contributing to the conservation of these magnificent insects.

What Will You Do? You'll start by learning about **milkweed** - the only home and food source for **monarch caterpillars**. You'll grow milkweed, observe how monarchs transform through their life stages, and even help caterpillars safely become butterflies! But that's not all; you'll play games, collect data, and make a wish during a special butterfly release event.

As you turn each page of this workbook, you'll discover new facts, engage in fun activities, and become a **monarch** expert. So, grab your magnifying glass, put on your explorer's hat, and let's begin this amazing adventure!

Understanding Milkweed
The Magical World of Milkweed: A Home for Monarchs and a Tale of Healing

A Plant Full of Wonders

Ready to dive into a world filled with tiny caterpillars and a plant that's like a superhero for them? This hero is none other than the milkweed! It's not just a plant; it's a cozy home and the only food that little **monarch** caterpillars will eat.

How Do We Name Our Plants?

Because plant scientists from around the world may speak different languages, we use scientific names for plants that are the same no matter what language you speak. These scientific names are usually in Greek or Latin. In this system, each plant is given two main names: the **genus** and the species.

How We Name Plants

SCAN THE QR CODE

- **Genus:** This is the first part of the plant's scientific name, like a family name. (like Bowers, Cox, or Alvarez for example) It groups together plants that are closely related. For example, all roses belong to the **genus** "Rosa."
- **Species**: This is the second part and it's more specific, like an individual's first name (like Lilah, Gracelyn, or Eden for example.). It tells us exactly which plant we're talking about within its genus. For example, in "Rosa canina," 'canina' specifies it's a particular type of rose.

Plant **species** names provide insight into their features, the region they are associated with, or they honor someone significant in their discovery or study. The **binomial nomenclature** system, which gives each **species** a unique Latin name, is used universally by scientists to avoid confusion that might arise from common names.

Apart from scientific names, we also have common names for plants. These are the names we use in everyday conversation in our own language. For instance, "Rosa canina" is commonly known as the dog rose (like Sissy, or Peanut). Common names are easier for most people to remember and use, but they can vary from place to place. That's why scientific names are so important in the scientific community, ensuring everyone is clear about which plant is being referred to, regardless of their native language.

Milkweed: Named After a Healer

The genus name is fun to learn about because there is usually an interesting mythological creature connected to it. The **milkweed** plant is scientifically named "Asclepias" after **Asclepius** because, just like him, it's super important for keeping things alive and healthy. For **monarch** butterflies, **milkweed** is essential. It helps the tiny caterpillars grow big and strong, just like **Asclepius** helped people long ago in his healing temples.

The different milkweed plants are in the genus **Asclepius**. So, every time you see a **milkweed** plant, remember **Asclepius**, the great healer, and think about how this humble plant is a superhero for the beautiful **monarch** butterflies! Let's learn the story of Asclepius so this isn't just a fancy word to memorize.

A Story from Long Ago

Let's start at the beginning. Where is that? We can start in ancient Greece. Let me tell you an enchanting story from the olden days. In ancient Greece, there was a healer named **Asclepius** (as-KLEE-pee-us), famous for his amazing ability to cure people and animals.

Asclepius's dad was **Apollo**, known for driving the sun across the sky in his chariot. His mom, Coronis, sadly couldn't be with him as he grew up. So, **Apollo** sent him to be raised by a wise **centaur** named Chiron. Imagine that – a creature that's half-horse and half-human!

Chiron was super smart, especially in healing. He taught **Asclepius** all his knowledge. **Asclepius** learned so well that he could even help people who were extremely sick, almost like a miracle worker!

One thing that made **Asclepius** famous was his staff with a snake wrapped around it. Today, this symbol represents doctors and medicine. In those old stories, snakes were seen as magical creatures that could heal themselves by shedding their skin.

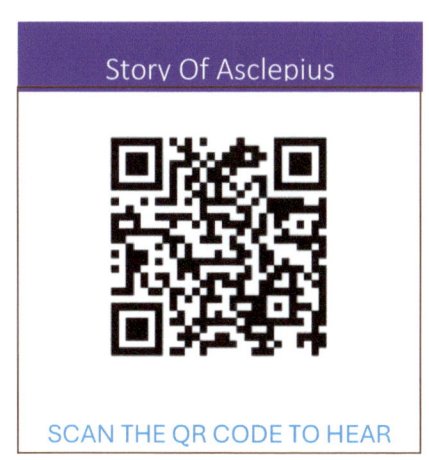

Story Of Asclepius

SCAN THE QR CODE TO HEAR

Asclepius became so famous for his healing powers that people built special places in his honor, called Asclepieia. These were like the hospitals of ancient times, where people went for healing.

Common Name "Milkweed"

Have you ever wondered how **milkweed** got its common name? It all starts with its distinctive sap and historical uses. The common name "milkweed" comes from the plant's milky white sap, which is visible when you break its stem or leaf. This sap contains latex, a mixture that was once explored as a potential source of rubber by American scientists during World War II. Additionally, the sap has been used in traditional medicine, though it's important to remember that it can be toxic if ingested. The name "butterfly weed," for the colorful Asclepias tuberosa, highlights its attractiveness to butterflies and its bright, ornate flowers that resemble the vibrant wings of these insects.

Types of Milkweeds

The various **species** of **milkweed** play a crucial role in the lives of **monarch** butterflies, and each **species** contributes in slightly different ways depending on its characteristics and the habitat it thrives in. Here's how these different kinds of **milkweed** support monarchs:

1. **Common Milkweed (Asclepias syriaca):** Found in many places, providing food for caterpillars.
2. **Butterfly Weed (Asclepias tuberosa):** Bright flowers attracting adult monarchs, great for gardens.
3. **Swamp Milkweed (Asclepias incarnata):** Thrives in wetlands, offers food and nectar for monarchs.
4. **Showy Milkweed (Asclepias speciosa):** Large flowers, key for monarchs in the western U.S.
5. **Prairie Milkweed (Asclepias sullivantii):** Important in prairies, similar to common milkweed.
6. **White Milkweed (Asclepias variegata):** Suited for forest areas, with unique white flowers.
7. **Green Milkweed (Asclepias viridis):** Adapts to dry areas and blooms early in the season.

Each **species** of **milkweed** contributes to the **monarch** population by offering caterpillars a vital food source and adult monarchs' **nectar** for energy. The diversity of **milkweed species** across different habitats and climates ensures that monarchs have access to resources throughout their range and migration routes.

Design a Pretend Milkweed

The **species** names of plants like **milkweed** can reflect either their physical characteristics or honor a botanist. For example, **speciosa** means beautiful. So, **Asclepias speciosa** has the most beautiful flowers.

"Did you know that many good fiction authors love to use their imagination to create new things in their stories, like plants, animals, and even whole worlds? When they make up names for these things, they often think about what makes them special or different. Just like how 'Asclepias speciosa' means a milkweed with beautiful flowers, authors invent names that give hints about their creations.

For example, if an author imagined a plant that glows in the dark, they might call it 'Asclepias luminosa', where 'luminosa' means 'light' or 'bright'. This way, the name itself gives us a clue about the plant!

Now, it's your turn. Think of a special characteristic or place for your milkweed plant. Is it super tall? Does it grow in cold places? Use your imagination to give it a species name that matches its unique feature or home. Remember, just like real scientists and authors, you can be creative and thoughtful in naming your Asclepias. Have fun creating your own magical milkweed world!"

Here are some pretend milkweeds that we could use in a story. Design your own milkweed plant. The species name should reflect the plant's characteristics or environment. **Asclepias** _____.

Make up your own pretend milkweed plant and draw a picture of it. You can use it in your story.

Asclepias Acer	**Asclepias Aquaticus**	**Asclepias Aureus**
Acer means "sharp."	Aquaticus means "living in water."	**Aureus** means "golden."

These are pretend milkweed types that you could use in a fiction story.

Reflection Questions:

1. What new things did you learn about milkweed and why it's important for **monarch** butterflies?
2. Can you think of other plants that are vital for specific animals? Why is this kind of relationship important in nature?
3. How does the story of **Asclepius** help you understand the importance of healing and care in nature?
4. If you could transform like a caterpillar, what would you want to become and why?
5. Which type of **milkweed** do you find most interesting, and what is unique about it?
6. How do different types of **milkweed** support the environment and **monarch** butterflies in various ways?

Glossary for Section 1: Understanding Milkweed

1. **Apollo (uh-POL-oh)**: A character from ancient myths, known for driving the sun's chariot across the sky.
2. **Asclepius (as-KLEE-pee-us)**: A healer from ancient myths, famous for his amazing healing powers.

3. **Binomial Nomenclature (bye-NOH-mee-uhl noh-men-KLAY-chur)**: A system for giving each **species** of plant and animal a unique two-part Latin name.
4. **Centaur (SEN-tawr)**: A creature from **myth** with the upper body of a human and the lower body of a horse.
5. **Genus (JEE-nus)**: The first part of a scientific name; groups closely related **species** together, like a family name.
6. **Milkweed (MILK-weed)**: A plant with milky sap, the exclusive food for **monarch** caterpillar.
7. **Monarch (MON-ark)**: A type of butterfly with a unique long migration and distinct orange and black wings.
8. **Myth (mith)**: A traditional story from ancient times, often explaining natural phenomena or cultural practices.
9. **Nectar (NEK-tur)**: A sweet fluid produced by plants, used by butterflies and other insects for food.
10. **Species (SPEE-sheez)**: The second, more specific part of a scientific name; identifies the exact organism within its genus.

Predators

SCAN THE QR CODE TO HEAR THESE READ

Who's After the Monarchs?

Hey, have you ever wondered what dangers **monarch** butterflies face in their world? While monarchs are famous for their beautiful wings and long migrations, they also must watch out for several **predators** that can make life tricky for them. Let's find out who these **predators** are and how they affect the monarchs.

The Feathered Foes

Some birds, like orioles and grosbeaks, have figured out how to eat monarchs without getting sick from the toxins in their bodies. These birds have learned to avoid the most toxic parts of the butterfly.

The Tiny Hunters

When monarchs are resting for the winter, certain mice, like the black-eared mouse, can sneak up and make a meal out of them. These mice don't seem to mind the toxins in the monarchs at all.

The Small but Mighty

You might not think of insects as being dangerous, but for a **monarch caterpillar** or butterfly, they can be big trouble. Wasps, ants, stink bugs, and even spiders can attack monarchs, especially when they are still eggs or **caterpillars**.

The Hidden Danger

There's a tiny creature called the **Ophryocystis elektroscirrha,** or OE for short, that can make monarchs very sick. It's a **parasite** that gets inside the **caterpillars** and can harm them as they grow.

The Stealthy Predators

Praying mantises are like the ninjas of the insect world. They are good at catching monarchs with their quick moves and strong front legs.

Even though these **predators** sound scary, they are a natural part of the environment. Monarchs have their own ways of staying safe, like the toxins they get from **milkweed** plants. But understanding these dangers helps us realize how amazing monarchs are for surviving and making their incredible journeys every year!

Fun Fact!

Did you know that the bright colors of monarchs are a warning to **predators**? They're saying, "I don't taste good!" This is called "**aposematism**," and it's a cool trick that many animals use to stay safe.

So, next time you see a **monarch** butterfly, remember that it's not just a pretty insect but also a survivor in a world full of challenges!

Activity:

Write your own "Monarch Rescue Adventure" Story

Highlight one option under each heading. Then write a story using those options. Feel free to illustrate your story! We used DALLE AI to illustrate our stories!

Choose Your Milkweed Garden:

- *Option 1:* A vast field of Common Milkweed.
- *Option 2:* A colorful garden with Butterfly Weed.
- *Option 3:* A wetland area filled with Swamp Milkweed.

Decision: Which one do you pick for your adventure's setting?

Select the Predator:

- *Option 1:* A hungry bird, eyeing the monarchs.
- *Option 2:* A sneaky spider weaving its web.

- *Option 3:* A group of ants marching towards the **monarch** eggs.

Decision: Which **predator** challenges your **monarch**?

Discover Your Superpower:

- *Option 1:* The power to grow plants instantly for hiding.
- *Option 2:* The ability to create wind gusts to sweep away **predators**.
- *Option 3:* A magical shield to protect the monarchs.

Decision: What power do you choose to defend the monarchs?

Plan Your Rescue:

- *Scenario:* The **predator** is closing in on the monarchs. How do you use your chosen superpower to save them?

Decision: Describe your heroic rescue plan.

The Monarch's Secret:

- *Twist:* The **monarch** you save reveals a secret. What is it? Maybe it's a lost prince or a hidden **monarch** kingdom! You decide.

Decision: Write what the **monarch** tells you.

Create the Happy Ending:

- *Closure:* How does your story end after the rescue?

Decision: Conclude your adventure with a happy ending.

The Magical Journey of Monarch Butterflies

Your Story

Lilah's Story

Eden's Story

Gracelyn's Story

Reflection Questions:

- What **predators** do **monarch** butterflies face, and how do they protect themselves?
- Why do you think it's important for us to understand the challenges monarchs face in the wild?

Glossary for Section 2: Predators

1. **Aposematism (ah-poh-SEM-uh-tiz-um)**: A way animals warn predators that they are not good to eat, often through bright colors or patterns.
2. **Asclepius (as-KLEE-pee-us):** A healer from ancient myths, famous for his amazing healing powers.
3. **Caterpillar (KAT-uh-pil-lur)**: The larval stage of a butterfly or moth; a stage in a butterfly's life when it looks like a small worm.
4. **Monarch (MON-ark)**: A type of butterfly with a unique long migration and distinct orange and black wings.
5. **Ophryocystis elektroscirrha (oh-fry-oh-SIS-tis e-lek-tro-SKIR-uh)**: (You can just say "OE"). A parasite that can infect monarch butterflies and caterpillars.
6. **Parasite (PAR-uh-site)**: An organism that lives on or in another organism (its host) and benefits by getting nutrients at the host's expense.
7. **Predator (PRED-uh-tur)**: An animal that naturally hunts and eats other animals.

Milkweed Safety

Safe Handling of Milkweed: A Must for Young Monarch Helpers!

Hey, intrepid nature detectives! While **milkweed** is a superstar in the world of **monarch** butterflies, it's super important to handle it with care. Ready to learn how to be a safe and smart explorer with this special plant?

Always Suit Up with Gloves

Did you know the **milkweed** sap, that sticky stuff inside the plant, can make your skin unhappy? So, whenever you're on a milkweed mission – planting, moving, or just checking it out – remember to wear gloves. This is like wearing armor; it keeps your hands safe and feeling good!

The "No-Tasting" Rule

Here's a funny but very serious rule: never, ever taste or eat **milkweed**! It's a fancy feast for **monarch caterpillars** but a big no-no for us humans. Eating **milkweed** can give you a tummy ache because of some powerful chemicals it contains.

The Super Soap Hand Wash

Post-milkweed adventure, it's crucial to wash your hands thoroughly with soap and water. This is like a magic spell to remove any sap that might have sneaked onto your skin, protecting you from any irritation.

Fun Fact Alert! 🦋

Guess what? The same stuff in **milkweed** that can bother our skin is like a superhero shield for monarch caterpillars! It protects them from creatures that might want to snack on them.

Milkweed: Your Garden's Butterfly Station

Got **milkweed** in your garden? Fabulous! You're a **monarch** hero! Just keep in mind our safety tips: gear up with gloves, remember the "no-tasting" rule, and always do the super soap hand wash. With these tricks up your sleeve, you can admire **milkweed's** beauty, cheer on the **monarchs,** and stay safe!

Activity: Milkweed Safety Pledge & Badge

Be a Milkweed Safety Ambassador!

Materials Needed:

- Paper or cardstock
- Coloring supplies (markers, crayons, or colored pencils)
- Safety Pledge Template (provided below)
- Badge Template (provided in the workbook)

Steps:

1. **Create Your Safety Badge**: Use the Badge Template for students to design their own 'Milkweed Safety Ambassador' badge. Encourage creativity!
2. **Sign the Pledge: Milkweed Safety Pledge**

MILKWEED SAFETY PLEDGE

I, _____, promise to be a Milkweed Safety Ambassador. I will always wear gloves when handling milkweed, never taste or eat the plant, and wash my hands thoroughly after touching it. I understand these rules keep both me and the monarch butterflies safe and healthy.

_____
Great Grandma Esarey
Monarch Specialist

_____
Grandma Esarey
Milkweed Specialist

The Magical Journey of Monarch Butterflies

Reflection Questions:

1. Why is it important to handle **milkweed** safely, and what precautions should you take?
2. How does knowing about the safety measures change the way you think about interacting with plants and nature?

Glossary for Section 3: Milkweed Safety

1. **Caterpillar (KAT-uh-pil-lur)**: The larval stage of a butterfly or moth; a stage in a butterfly's life when it looks like a small worm.

2. **Milkweed (MILK-weed)**: A plant with milky sap, the exclusive food for monarch caterpillar.

3. **Monarch (MON-ark)**: A type of butterfly with a unique long migration and distinct orange and black wings.

4. **Predator (PRED-uh-tur)**: An animal that naturally hunts and eats other animals.

Parts of the Milkweed Plant

The Important Parts of a Milkweed Plant

Milkweed plants are like little worlds full of wonders, especially for **monarch** butterflies. Every single part of the **milkweed** plant has a special job that helps both the plant and the **monarchs**. Let's take a closer look at what each part does:

Leaves: *The Caterpillar's Cafeteria*

The leaves are like a non-stop buffet for monarch **caterpillars**. They munch on these leaves day and night, getting all the nutrients they need to grow. The leaves are super important because they're the only thing **monarch caterpillars** eat!

Flowers: *Nature's Perfume Advertisements*

Milkweed flowers are like bright ads, saying "Hey, over here!" to **pollinators** like bees and butterflies. They come for the **nectar** and help the plant by moving pollen from one flower to another. Later, these flowers turn into seed pods.

Stems: *The Plant's Backbone*

Think of the stems like the plant's skeleton. They hold up the milkweed and keep it sturdy. The stems also have special sap inside them. This sap is filled with nutrients, and guess what? It's this sap that gives **monarchs** their "don't-eat-me" shield against **predators**!

Roots: *The Underground Network*

The roots are busy working under the soil, even though we can't see them. They drink up water and minerals from the ground and send them up to the rest of the plant. Without roots, the **milkweed** couldn't get the food and drink it needed to grow.

Pods: *Parachutes for Seeds*

After the flowers are done blooming, **milkweed** forms pods. These pods are like little houses for the seeds. When they open, the seeds come out on silky threads, almost like parachutes. These parachutes let the seeds float away on the wind to grow new milkweed plants somewhere else.

Reflection Questions:

- Which part of the **milkweed** plant do you think is most important for **monarch** butterflies, and why?
- How do different parts of the plant support not only the **monarchs** but the **ecosystem** as a whole?

The Magical Journey of Monarch Butterflies

Challenge

Instructions: Below are 16 words, all scrambled together. Your task is to sort them into four groups of four words. Each group has a common theme. Once you've sorted them, write down the theme for each group! *(The first one is done for you to get you started.)* Scan QR codes for hints.

Instructions

Words to Sort:

Birds	pods	Apollo	spider
leaves	caterpillar	centaur	stem
Asclepius	roots	chrysalis	miracle healing
butterfly	egg	Fire ants	mice

1. _____

2. _____

3. _____

4. _____

Time to Be a Plant Scientist with Milkweed!

Hey, young scientists! Growing **milkweed** is more than just planting and watching. It's about becoming a real plant scientist! Here's a cool experiment to watch your **milkweed** grow in a special way:

What You'll Do:
- **Make Your Milkweed Bushy**: We're going to make the **milkweed** grow wide and bushy, not just tall. This is called "pinching back" or "pruning."

How It Works:
- **The Science of Snipping**: When you carefully pinch off the top parts of the some plants, it starts growing more stems and leaves from the sides. It's like telling the plant, "Hey, let's grow wider, not just taller!" We don't know if this works for **milkweed** or not. This doesn't work for all plants.
- **Why Do We Care?**: Making a plant bushy would be great for **monarch caterpillars**—they would get more space and more leaves to munch on! Plus, it would make the plants look full and pretty in your garden.

By doing this experiment, you'll see how plants respond to your care and learn about making a happy, healthy home for monarch caterpillars! Let's get our gloves on and start our plant science adventure! 🌱 🦋

Experiment: The Effect of Pinching on Milkweed Plant Growth

Experiment: Milkweed Growth and Branching

Objective: To determine how pinching (pruning) the tips of milkweed affects plant branching and foliage growth.

Hypothesis

Form a hypothesis about whether pinching milkweed plants encourages more branching and foliage growth compared to unpruned plants.

Materials

- At least 10 milkweed plants of the same species and similar size
- Gardening gloves
- Pruning shears
- Ruler or measuring tape
- Labels or markers
- Notebook for observations

Procedure

1. **Prepare Your Plants:**
   - Divide the milkweed plants into two groups: Group A (Control) and Group B (Experimental).
   - Label each plant clearly.
2. **Initial Measurements:**
   - Measure the initial height and count the number of branches on each plant. Record in the notebook.
3. **Pinching Process:**
   - For Group B, pinch or cut off the top set of leaves or the growing tip of each plant.
4. **Observations:**
   - Over a period of 4-6 weeks, regularly measure the height and count new branches on each plant.
   - Record any changes in foliage, health, and overall growth.
5. **Care and Maintenance:**
   - Ensure both groups receive equal amounts of water, sunlight, and other care.

Variables

- **Independent Variable:** Pinching the plants (Group B only).
- **Dependent Variable:** Growth of the plants (height, number of branches, foliage).
- **Control Variables:** Species of milkweed, soil type, water, sunlight, care.

Data Analysis

- Compare the average growth and branching between the two groups.
- Use graphs or charts to visualize the growth patterns.

Conclusion

- Determine if pinching positively affects the branching and foliage growth of milkweed.
- How this knowledge can be applied in gardening or butterfly habitats?

Safety Note

Always wear gloves when handling milkweed due to the potentially irritating sap.

This experiment introduces students to basic scientific methods and the concept of plant growth manipulation. It's a practical way to explore botany and understand how human actions can influence plant development.

Experiment Reflection:

- What did you learn about plant growth?
- Why was it important that plants in both groups had the same sun, water, etc.?
- How can your findings be applied in gardening or agriculture?

Glossary for Section 4: Parts of the Milkweed Plant

1. **Caterpillar (KAT-uh-pil-lur):** The larval stage of a butterfly or moth; a stage in a butterfly's life when it looks like a small worm.
2. **Ecosystem (EE-koh-sis-tuhm):** A community of living things and their environment, working together as a system
3. **Milkweed (MILK-weed):** A plant with milky sap, the exclusive food for monarch caterpillar.
4. **Monarch (MON-ark):** A type of butterfly with a unique long migration and distinct orange and black wings.
5. **Nectar (NEK-tur):** A sweet fluid produced by plants, used by butterflies and other insects for food.
6. **Pollinator (POL-li-nay-tur):** An animal that moves pollen from one flower to another, helping plants to produce fruits and seeds.

7. **Predator (PRED-uh-tur):** An animal that naturally hunts and eats other animals.

The Amazing Lifecycle of a Monarch Caterpillar

The Journey from Egg to Butterfly

Monarch butterflies go through a remarkable life cycle, transforming from an egg to a **caterpillar**, into a **chrysalis**, and finally emerging as a stunning butterfly. But there's more! The **caterpillar** stage itself is divided into mini stages called "**instars**". Let's explore these stages and the incredible process of **metamorphosis**!

Understanding Metamorphosis

Before we delve into **instars**, let's talk about **metamorphosis**. This is the process where the **caterpillar** completely transforms its body and becomes a butterfly. Think of it as a caterpillar going into a sleeping bag (the **chrysalis**) and emerging as a completely different creature. It's nature's magic trick!

What's an Instar?

Imagine outgrowing your clothes every few days. That's kind of what happens to a **monarch caterpillar**! As it grows, it gets too big for its skin and needs to "change" into a bigger one. These changes are called **instars**.

1. **First Instar:** The Tiny Beginnings
    - When the **caterpillar** first hatches, it's in its first instar. It's super tiny, about the size of a small ant, and starts munching on **milkweed** leaves right away.
2. **Second Instar:** Getting Stripy
    - Soon, the **caterpillar** starts showing its famous stripes. It's still small but is getting bigger every day!
3. **Third Instar:** Growing Bigger
    - Now, the **caterpillar** is getting more noticeable with bright stripes. It's eating a lot and growing fast.
4. **Fourth Instar:** Almost There!
    - The **caterpillar** is much bigger now. It has long tentacles on its head and back, and it looks more like the **caterpillars** you're used to seeing.
5. **Fifth Instar**: The Final Stage

- This is the last stage before it turns into a **chrysalis**. The **caterpillar** is the biggest it will ever be and its stripes are very clear. It's getting ready for a big change!

Eating, Growing, Changing

During each **instar** stage, the **caterpillar** eats and grows a lot. It's busy gathering energy for the next big step: turning into a **chrysalis**.

From Caterpillar to Chrysalis

After the fifth **instar**, the **caterpillar** finds the perfect spot to start its transformation. It's an exciting time because soon, it will become a **chrysalis** and then, a beautiful **monarch** butterfly!

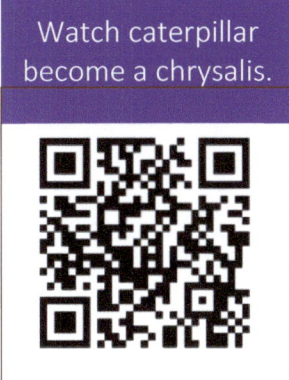
Watch caterpillar become a chrysalis.

Fun Fact!

Did you know that **caterpillars** can increase their body size more than 2,000 times while they are growing? That's like a baby growing to be the size of an elephant!

The Magical Transformation: Metamorphosis in Mythology

A Story of Change

How have people understood something as magical as "metamorphosis" throughout time? Metamorphosis is a way to talk about an amazing change. The word comes from an old language called Greek, where "meta" means change and "morphosis" means form. It's like going from a caterpillar to a beautiful butterfly!

Myths and Magical Beings

In ancient stories, especially in a big collection of myths called "**Metamorphoses**" by a guy named Ovid, **metamorphosis** was a super popular theme. Magical beings in these stories had the power to transform themselves or others into different forms. Imagine being able to turn into a tree or a bird whenever you wanted!

One of these tales is about a nymph named Daphne. She loved her freedom so much. One day, a sun chariot driver named **Apollo** really liked her, but she wasn't interested. To escape him, she turned into a laurel tree! Can you believe that? Instead of being sad, Apollo made the laurel tree a special symbol to remember her. (You remember **Apollo**! He was **Asclepius's** dad.) **Apollo** then started wearing a crown made of laurel leaves to keep her near him.

Why Does This Matter?

These old myths tell us about how people long ago thought about the world. They tried to make sense of changes in nature and life by telling stories about magical beings who could transform. Just like how a caterpillar transforms into a butterfly, these stories showed transformation as something mysterious and amazing.

Our Own Transformations

Even though we can't turn into trees or animals, we all go through changes, like growing up. Maybe next time you see a butterfly, you'll remember these cool stories and think about all the changes waiting for you!

Activity

Rewrite your story from Section 2. Add an event where a metamorphosis occurs to help in the rescue.

Reflection Questions:

- What part of the **monarch**'s lifecycle do you find most fascinating, and why?
- How does understanding the **monarch**'s lifecycle help us appreciate these butterflies more?

Glossary for Section 5: The Amazing Lifecycle of a Monarch Caterpillar

1. **Caterpillar (KAT-uh-pil-lur)**: The larval stage of a butterfly or moth; a stage in a butterfly's life when it looks like a small worm.
2. **Chrysalis (KRIS-uh-lis):** The pupa stage of a butterfly; the stage where the caterpillar turns into a butterfly.
3. **Instar (IN-star):** A stage between two molts in the development of a caterpillar.
4. **Metamorphosis (met-uh-MOR-foh-sis):** The process of transformation from an immature form to an adult form in animals like butterflies.
5. **Milkweed (MILK-weed):** A plant with milky sap, the exclusive food for monarch caterpillar.
6. **Monarch (MON-ark):** A type of butterfly with a unique long migration and distinct orange and black wings.

Identifying the Gender of Monarch Butterflies

Identifying the gender of monarch butterflies can enhance our understanding of their life cycle, behavior, and the specific roles they play within their habitats. This knowledge is particularly useful when observing their interaction with milkweed plants.

Male Monarch Butterflies

Male Monarchs can be distinguished by their vibrant colors and distinctive physical markers. The most notable feature is the presence of two small, black spots on the inside of their hind wings, which are absent in females. These spots are scent glands used during mating. Additionally, the veins on a male Monarch's wings are slightly thinner compared to those of a female, giving the males a somewhat more delicate appearance.

Male (Drawing by Eden Alvarez)

Female Monarch Butterflies

Female Monarchs, while sharing the iconic orange and black patterning of their male counterparts, lack the black scent glands on their wings. Their wing veins are thicker and more pronounced, which is a key difference from males. Females play a crucial role in the continuation of the species, laying eggs on milkweed plants to ensure the next generation of Monarchs.

Female:

Drawing by Eden Alvarez

Reflective Questions:

1. Why do you think it's important to know if a Monarch butterfly is male or female when observing them on milkweed plants?
2. How do you think identifying the gender of Monarch butterflies could help us take better care of them and their environment?

A Fluttering Legacy: The Monarch's Gift from Great Grandma Carol

In the heart of a lush garden, alive with the melodies of nature and the sweet aroma of milkweed in bloom, three young girls, Lilah, Eden, and Gracelyn, stood with anticipation. They were not just any girls; they had become budding nature explorers and scientists, thanks to the influence of Lilah's remarkable Great Grandma Carol. A passionate naturist herself, Great Grandma Carol had not only ignited in them a fervent love for monarch butterflies but had also opened the doors to a vast world of nature, science, and storytelling.

Great Grandma Carol, with her deep understanding of the natural world, had a magical way of making every leaf, every creature, seem like a part of a grand story. From her, the girls learned not just about the transformation of monarchs but also about the interconnectedness of ecosystems, the importance of conservation, and the joy of discovery. She had a special talent for weaving tales that combined myth, science, and observation, turning their learning journey into an adventure.

Today, Lilah, Eden, and Gracelyn were about to release the butterflies they had nurtured from caterpillars, a ritual that symbolized much more than just a goodbye. Their grandmother, following in the footsteps of Great Grandma Carol, reminded them gently, "Make a wish."

Holding the delicate monarchs carefully, each girl made her wish. Their words were imbued with the hopes for safe travels for their winged friends and a deeper aspiration for a world where nature and humans exist in harmony – a dream first kindled by Great Grandma Carol.

As the monarchs fluttered away from the tent, their wings glinting in the sunlight, the garden seemed to resonate with the spirit of exploration and learning. This wasn't just a release; it was a celebration of the continuous cycle of life and the curiosity and care that Great Grandma Carol had fostered in them.

This moment was about much more than the transformation of caterpillars into butterflies; it marked the growth of three young girls into guardians of nature, armed with a thirst for knowledge and a heart full of stories. The experience was a living testament to Great Grandma Carol's legacy, a reminder of the beauty, resilience, and wonders that await in the natural world for those eager to explore and learn.

www.ingramcontent.com/pod-product-compliance
Lightning Source LLC
Chambersburg PA
CBHW040448220526
45473CB00004B/1559